DEWEY'S INSTRUMENT TROUBLESHOOTING HANDBOOK

By Ralph Dewey

Glen Enterprises
PO Box 1201
Deer Park, TX 77536-1201

DISCLAIMER:

* This book is written for the reader/user who has at least a general knowledge of instrumentation, controls and electrical systems. Anyone with a minimal understanding of instrumentation or none at all, should not use the ideas in this book.
* All of the tips, tricks, designs, methods and ideas are general in nature and may or may not fit your specific situation.
* Before using any of the ideas in this book, obtain the approval of a competent instrument engineer.
* Neither the author or the publisher shall be held responsible for information errors or any applications found in this book, or any losses incurred from the use of the information.
* Always use safe practices and procedures when working with electrical and electronic equipment and instruments.

Dewey

ISBN 1-880215-30-6

TROUBLESHOOTING MIND SET

$\mathcal{P}$ Before any problem can be solved, you must have a good understanding of how the controls or equipment are supposed to function and operate. Depending upon the situation, you should be able to find out how it operates by reading the manufacturer's literature, by looking at drawings, by asking experienced operators, or by tracing out the wiring or system. It may take a combination of all of these ideas. If nothing else, make a sketch of all the pieces and components of the control system. Then go over it until you understand how it should work.

$\mathcal{P}$ Most troubleshooting efforts use three basic steps.

❶ Obtain a good understanding of the symptoms and how the instrument or system is not functioning properly.

❷ Investigate to determine why the instrument or system is not functioning properly. Look for the root cause. Often you will need to use your experience to come up with possible scenarios that could cause the symptoms.

❸ Determine what needs to be done to resolve the problem and then take corrective action.

$\mathcal{P}$ Most instrument and electrical problems fall into one of three categories, totally not working, poor signal quality or intermittent.

① The easiest problems to resolve are those where a voltage is missing, the power is off or some feature doesn't function at all. For a missing voltage or power lose, check the power source, fuses, wiring or point where it fails to appear. For nonfunctioning features, you may need to be able to track backwards until you see why it doesn't work.

② Signals with poor quality can be difficult to uncover. Often oscilloscopes or other sophisticated test equipment is required for troubleshooting. Try to isolate or determine which part of the circuit that can generate poor signal quality.

③ Intermittent problems are by far the most difficult to resolve. For more information, see the section called, "Troubleshooting Intermittent Problems" for some ideas on how to approach problems that only appear occasionally.

🔎 When troubleshooting, one question you need to answer is whether the symptom is a problem or a result? For example, if you discover that a fuse has blown, that's probably not the problem, it's more likely the results of a problem. Something has caused the fuse to blow. And most likely, if you simply replace the fuse, it will blow again soon. Keep looking for what caused the blown fuse. Look for the root cause of the problem.

🔑 Strive to make a long-term repair, rather than just make a quick fix. If you discover that the tap on a pressure transmitter is plugged, that there is condensate in the dry leg of a level transmitter or that mud daubers have plugged an exhaust port of a solenoid, take steps to keep it from reoccurring. Often, you can relocate an instrument, add a seal leg, install a filter or choose a different measurement technology to keep the problem from coming back.

PROACTIVE MAINTENANCE

There is always a reason why an instrument fails. A proactive craftsperson looks for ways to keep it from reoccurring. Here are the steps to follow.

1. Determine the root cause of the failure.

2. If possible, find a way to eliminate the problem.

3. If the problem can not be eliminated, reduce or slow down the problem.

4. If the problem can not be reduced or slowed, make provisions to make maintenance easier next time.

Let's take an orifice flow meter for example. Each time the meter fails, solids are found packed in the process taps. You find out that the fluid has suspended solids and that tend to plug the orifice taps. (Step 1, You've found the root cause.) Is there a way to keep the suspended solids out of the process? It might be possible for manufacturing to make changes to their process that would solve the problem. (Step 2, Try to eliminate the problem.) Can the flow meter be oriented differently? Could the orifice be changed to a wedge meter which resists plugging? Could liquid purges be used to keep the taps clean? Could D/P cells with remote capsules be used? Could the taps on the orifice flange be rotated so solids do not gravitate down into them? (Step 3, Try to reduce or slow down the problem.) All of the previous suggestions may not be workable or be possible to implement. You may have budget limitations. So your last proactive option, step 4, is to try to make repairs easier next time. One idea is to permanently install rod-out devices on the orifice taps. Every time you get a work order to repair the plugged flow meter, you can easily clean out the taps.

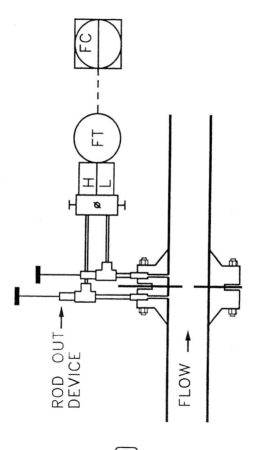

FC

FT

H L

ø

ROD OUT
DEVICE

FLOW →

7

🔑 One of the best maintenance ideas that I have seen was the brainchild of an area maintenance supervisor. His unit was known for being tough on control valves. At each annual turnaround (which lasted about five days) there were about 50 control valves that had to be rebuilt. With that large amount control valves being sent to the central instrument shop at one time, it made it difficult to repair them in the short period allocated for the turnaround. He had a solution. Over a period of several years (to spread out the cost), he slowly bought a duplicate of each control valve that needed to be rebuilt. He had a storage building where he stored the replacement valves. He tagged them with an "A" or "B" suffix to be able to keep track of them easier. The day before the turnaround, he positioned the replacement valve next to the ones needing to be rebuilt. In a matter of hours all of the valves could be swapped. The same craftsperson took out the bad valve and replaced it with the good valve. The turnaround took a lot less time. Afterwards, he would send in about 5 control valves to the central shop for overhauling every couple of weeks. It was a workload that they could easily handle. In a few months, all of the valves were rebuilt and he was ready for the next turnaround.

TECH TIP: You can test a control valve to determine if the trim is too small or too large. Perform the test with the controller in automatic and the process operating under normal conditions. Aim for a controller output of about 60% and settled out. If it requires the opening of the bypass valve, the trim is too small. If the inlet valve must be pinched down, then the trim is too large.

Another proactive idea is to head off problems before they get critical. One example of this concept dealt with pH meters. One of the daily routine chores was to check all of the pH instruments for accuracy. It took about half a day for one person, if done correctly. But it was so mundane that none of the craftspeople wanted to do it. The schedule originally called for one person to do it all week. The following week another person did it. With six people, it only came around once a month and a half. However, some craftsmen thought it was too boring to do for an entire week, so they decided to rotate it every day.

The job required grabbing samples to be taken at each pH meter and compared against a test pH instrument. The test pH instrument was calibrated against buffer solutions to ensure that it was reading correctly. If an instrument was found to be off, it was adjusted to match the value of the test pH meter. Since most of the pH instruments were in hot and harsh service, the sensors would start to deteriorate and eventually fail sometimes in a few weeks and sometimes in a few months. When craftspeople were rotated daily, they had more unexpected sensor failures in the middle of the night than with the weekly schedule. Here is the reason why. When a craftsperson noticed a drift on one of the pH instruments, he would make an adjustment. The next day a second person noticed the same drift and he too made an adjustment. But since it was a different person each day, the trend was not noticed. Finally at some point during the night, the sensor would totally quit or drift severely. This forced the plant to operate blindly until a craftsperson could be called out to repair it. Originally when the schedule called for one person to check them all week, that person would notice that he had to make adjustments to the same instrument several days in a row. He would notice the sensor was deteriorating and trending toward a failure. He would replace it with a new one before it failed. Catching instrument failures before they happen is always better than being surprised by them.

𝒫 Using two pH meters has some merit. I remember one unit where the probes easily became coated and would quit working. The process was tough on probes. You could readily tell when the pH probe was getting coated because the normal quick-changing PV (process variable) oscillations on the circular chart slowed down to a lazy sine wave. For years the solution was to replace the probe or clean it off. However, when the pH probe coated in the middle of the night, a craftsperson had to be called out to replace the probe. This left operations without a good way to view the process pH for several hours. To help resolve this problem a second pH instrument was installed. Both pH meters had probes in the process. Either one could be switched to show on the circular chart and be the input signal to the DCS. (By the way, you can't just switch the pH probes, because each probe must be calibrated to it's own instrument.) When the selected pH probe would start fouling, operations would flip the switch to the backup meter and keep functioning without a bobble. This idea all but eliminated the need for late night call outs by craftsmen and greatly improved the instrument's reliability. The news about having dual transmitters spread to another unit in the plant that was also having trouble keeping a reliable pH measurement. They were having problems with a centrifuge and some other process equipment. They discovered that if they kept the pH of the water precisely at 8.5 pH, calcium would not form on the equipment. They decided to install dual pH meters too. This gave manufacturing a consistent and reliable pH reading, which in turn allowed them to maintain their pH control at the 8.5 level.

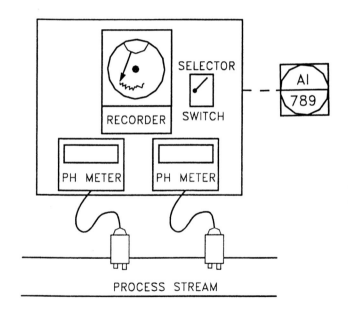

TROUBLESHOOTING APPROACHES

VISUAL CHECKS: Very often a simple visual check will reveal the problem. It is usually the very first troubleshooting method attempted. Look for obvious problems; broken wires, moisture, corrosion, plugged measurement taps, blocked process valves, a charred circuit board, wires crossed, broken springs, jumpers in the wrong position, wrong cam or parts installed incorrectly. One trick that may help is to compare the faulty device against an identical instrument in the same service.

SUBSTITUTION: Some problems can easily be found by replacing or substituting parts, circuit boards or components. Don't overlook the possibility of taking parts from a good unit temporarily. Trade out one part at a time so that you will know for sure which item was at fault. Here is a word of caution before substituting a part. Be sure that you can live with the consequences if that part gets "smoked" (damaged) too. If an electrical fault is ruining components, it might destroy the very last one in the plant. Do not blindly replace the same suspected part repeatedly. As a rule of thumb, don't swap out the same part more than three times. Think about it, surely the fault has to be somewhere else!

SHOTGUN METHOD: This is an expanded form of the "Substitution" method. Usually a company wants a problem fixed quickly and cheaply. But having both is not always possible. One of the fastest ways to find a problem is to replace every circuit card, cable, relay or plug-in device in the entire instrument at one time. Often it will get the equipment working quickly, but you seldom have a clue as to which part was the offending component. You may be trading expedience for a high maintenance cost. And what do you do with all those parts? Which one was bad? Or could you have more than one bad part? You might have to send all of them in to the factory to be checked? It's interesting that one PLC company reports that 80% of all bad cards sent to them for repair turn out to be perfectly good.

SIMULATION: Simulations that accurately imitate process or field conditions can help you determine if the instrumentation is functioning properly. Be sure that your simulation is valid and truly represents actual conditions. Otherwise your testing may not reveal the real root cause of the problem.

TEMPORARY JUMPERS: Temporarily jumping out switches or contacts may help you find or isolate an interlock problem that is keeping a piece of equipment down. Before jumping things out, consider all of the ramifications to be sure that it can be done safely.

DIVIDE & CONQUER: When you have a loss of voltage or a signal, find a mid-point (divide the signal path) and test for it. If you find it there, divide the circuit and test for it again down stream. If you didn't find it, divide and test for it upstream. Repeat this method until you narrow down the root cause of the missing voltage or signal. The same general troubleshooting method can be used when signal quality is low.

ELIMINATION: Some instrument problems could originate from the process, the installation, the electronics, EMI interference or equipment failure. Instead of proving which parts or components are bad, prove which ones are good. Employ methods, such as simulation, substitution or jumpers that can prove each part is not the problem. The process of elimination usually reveals the true problem.

CONSULTATION: There are times when difficult problems can easily be fixed if you ask the right person for advice. Often you can find old timers, operators, engineers, maintenance people or other experienced personnel who have encountered the same problem before. Use their experience to give you a direction to investigate. It's kind of silly to work on a problem all day and then discover that if you had asked, someone could have told you how to fix it

FACTORY HELP: Many manufacturers or factory representatives have troubleshooting guides in their literature to help you. Free factory help is usually a phone call away. They often have resident experts or technical people who can give you some tips and tricks to try. Look to see if the manufacturer has a troubleshooting guide on their web site.

INSTALL HELPS: Troubleshooting can be made a lot easier if you install gauges, meters, test points, relays with indicating neon bulbs, recorders or other devices that tell you what the instrument or process is doing. The more "eyes and ears" that you have, the more information that you have to help you solve the problem.

SET TRAPS: When you suspect that humans are the cause of problems or that someone is tampering with equipment, you can install tattletale devices. You can secretly tape a hair across the door of a cabinet or panel. If the door gets opened, you will know because the hair will be broken or disturbed. You can get special locks from a locksmith that will trap unauthorized keys. If you suspect that the process pressure goes beyond the normal parameters, you can install special gauge that has a follower pointer. Even though the process pressure returns to normal, the follower pointer will remain at the highest point.

FRESH START: After fighting a difficult problem for a long time, step back and look at it with new eyes. Think of all of the things that could cause those symptoms. Check all of it again as if it were a new problem. Don't assume anything is correct, even your test equipment. Don't overlook the fact that some problems are process related. By the way, if another technician has worked on the problem, they may have installed a wrong part or changed some settings. Try quick checks (the usual tests and the easy stuff to check) and if that doesn't reveal the problem, regroup and check things out thoroughly one at a time. By the way, if a new problem shows up while you are trying to resolve the original problem, you may be the culprit. Retrace your steps and thoroughly check everything that you have replaced, adjusted or touched. It is possible that you introduced a new problem and weren't even aware of it.

INTERMITTENT: Intermittent electronic problems that are heat related may be found by using a hot air gun to warm the circuit board. Use care not to overheat the components. Cooling a circuit board by spraying it with instant contact cleaner can sometimes cause intermittent problems to show up. Carefully flexing circuit boards may also reveal problem areas.

DUPLICATE ADDRESSES: Certain types of instruments monitored by a supervisory computer can have problems when modules or cards have duplicate addresses. One method for finding the module with the duplicate address is to systematically delete each individual address until proper communications with the computer is restored. Once the duplicate is found, all of the deleted addresses will have to be added back to the computer.

CONTROL VALVE TIPS

VALVE LEAKAGE TEST: One simple method to determine the approximate amount of leakage in a control valve is to cover the outlet connection with a wet paper towel. The water will allow the paper to adhere and form a seal. With the actuator in the closed position, apply about 20 PSIG of air to the inlet port. With zero leakage, the wet paper towel will maintain its seal. Small amounts of leakage will bubble up and break the seal occasionally. When you have excessive leakage, you will not be able to keep the seal formed very long.

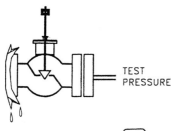

TEST
PRESSURE

BENCH SET: Most control valves have a standard bench set of 3-15 PSIG, but you find 6-30 PSIG too. When process differential pressures act upon a valve plug, they either tend to help the valve to open or to close. When a small actuator is chosen, often for economic reasons, its bench set may have to be non-standard in order to compensate for the process pressures.

7-15 PSIG AIR TO OPEN BENCH SET: When process conditions force you to purchase valves with A bench set such as 7-15 PSIG, the valve will have a non-standard bench set. In the case of an air to open valve, process pressures will tend to force the valve plug open. The non-standard bench set compensates for the high differential pressure across the valve trim. On the bench, the valve plug will begin to open when slightly more than 7 PSIG is applied to the actuator. When such a valve is placed in the actual process, the valve plug will begin to open with 3 PSIG applied. If you do not wish to deal with odd non-standard bench sets and desire to have a 3-15 PSIG bench set, you must purchase a larger actuator that can exert more force.

THINGS TO CHECK

CORRECT SOFTWARE VERSION: Sometimes computer or Local Area Networks (LAN) problems force you to re-install or download programs. If the master copy is an older version and your files are from a newer version, sometimes there will be an incompatibility. Check the revision or version number. Sometimes software companies will set up web sites to inform you of glitches. They may even let you download new software to replace your old versions. The year 2000 date problem may cause you problems that may not show up except on rare occasions.

When the Local controller will not maintain a set point.
If the process is oscillating, the controller may need tuning. Before you begin tuning, write down the current tuning parameters (values of gain, reset & rate). This will allow you to go back to the original settings ("as found") if tuning was not the solution. Check the control valve for a smooth stroke. A sticking valve stem can make it difficult to maintain a set point. Make sure that the controller has

the proper action. If process conditions have changed, some set points may not be achievable. Suppose that one of the two pumps in the system shuts down. Because the pump capacity has diminished, higher pressures or flows may not be possible. Is the valve positioner (if you have one) getting full air supply? If the controller is an old pneumatic model, one of the links my have become disconnected. Instability can be caused if a control valve is being operated too near the closed position. Many control valves become unstable below about 5% of their travel. Offset may be your problem. Offset is the steady state difference between the set point of a controller and the process variable. You may not have enough reset action. The reset mode was designed to eliminate offset. Four things can cause offset. Your controller could be out of alignment. However, that would not be very likely for DCS or PLC systems. Your tuning parameters may not be set for enough reset action. The set point indicator or process indicators may be out of calibration. Or it could be that something in the process will not allow the process variable to reach the set point. For example, if a pump has tripped out, the process variable may not be possible to achieve a desired flow rate. Another thing to check in electronic loops is for too much loop resistance. Most 24 VDC, 4-20ma loops can not have more than approximately 600 Ω. Too much loop resistance will limit the upper range of the transmitter signal (PV).

When the control valve will not function properly.
Test to see that the control valve will fully stroke under
the process conditions. There are times, when process
pressures will prevent a valve from seating off properly.
Some positioners have a bypass handle or switch. If the
position is reverse acting, split-range, or used for a special
function, it should not be bypassed. Has someone
bypassed it? Water in the air supply line feeding a valve
positioner will keep the valve from functioning properly.
To avoid water in the air supply, ensure that the air supply
tap is on the top of the supply header instead of on the
bottom.

**When the pressure transmitter is not sensing the
true process pressure.**
A plugged pressure tap is one of the most common
problems. You may consider installing a "rod out" device
so that the tap can be safely cleaned out periodically.
Some cases may need a purge on the process taps to
keep them open. In severe cases, you can install sealed
diaphragm sensors or special radial pressure spool pieces
that resist plugging.

**When the temperature transmitter is not sensing
the true process temperature.**
The sensor may not be bottomed out in its thermowell. If
the sensor is a thermocouple, the leads could be shorted
out in the connection head. (Temperature transmitters
will read the temperature where they are shorted instead
of at the sensor.) You might have product buildup

covering the thermowell. The physical location of your sensor may be giving you an erroneous reading or a large dead time. You might have the thermocouple wires connected backwards or not tightened correctly. You may be using the wrong thermocouple type. You could have a different range on the transmitter than the DCS. You might have moisture on the terminations or a partial short of the sensor wires.

When the orifice flow transmitter is not sensing the true process flow.

A plugged tap is a possibility. Does it need a purge or "rod out" device? Is your 3-valve manifold leaking by? An orifice plate installed backwards would show less flow than normal. You could have deposits or trash on the plate. Is there another way to verify the flow? Can the change in tank level be compared to the questionable flow rate?

When the D/P type liquid level transmitters is not sensing the true process level.

Plugged taps and lost seal legs are the two most common problems. One point of confusion is when you try to compare a tank's level by viewing the level glass versus the level transmitter when the process taps are physically at different heights. One of the drawbacks of static head level measurement is its vulnerability to changes in product density. Density fluctuations caused by either product or temperature variations, will effect the measurement accuracy.

Computer Re-boot:

Many computer bugs, glitches and problems are not solvable by the technician or maintenance personnel because of the scope and complexity of its programming. Factory help may be needed. However, many times a simple re-boot will restore the computer system. Its amazing how many times the trick of turning off the computer, waiting a about 10 seconds and then turning it back on has resolved problems.

Miscellaneous problems:

Remote level transmitters with capillary filled legs are often seen as a solution for tricky level problems. However, it's not common knowledge that they respond very slowly to process changes. In come cases, it may take four or five minutes to see a large level change. Therefore, quick-changing levels should not use this technology.

TROUBLESHOOTING CASE HISTORIES:
CASE #1

THE SYMPTOMS: A steam boiler was having problems during startup because the fuel gas pressure would swing, causing it to trip out on either high or low pressure. It always took many attempts to get the boiler ignited. No amount of tuning the controller seemed to help steady out the pressure fluctuations during a startup.

THE PROBLEM: There was too much lag time and gas volume between the sensor and the control valve. The control valve was located approximately 60 feet upstream of the pressure sensor. The fuel gas supply line was 3 inches in diameter. The time lag was tremendous making it difficult for the controller to maintain a given pressure. When the fuel gas pressure dipped too low, the controller started increasing and asking the control valve to open. By the time the pressure traveled down the pipe and the controller recognized that it was now too high, several moments passed before the controller could react in the opposite direction.

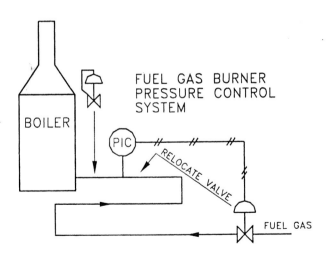

FUEL GAS BURNER
PRESSURE CONTROL
SYSTEM

BOILER

PIC

RELOCATE VALVE

FUEL GAS

THE CURE: The best solution would be to relocate the control valve right next to the sensor and the pressure controller to shorten the lag time. Because of cost considerations, a pressure regulator was installed downstream of the pressure sensor. It helped, but it was not a true resolution of the problem.

CASE #2
THE SYMPTOMS: A liquid level transmitter was mounted on the top of a vented tank. It seemed to track the level near the bottom of the tank but not at the upper levels.
THE PROBLEM: The level transmitter was a capacitance type. It had been installed in a stilling well from the top of the tank. No vent holes had been installed near the top of the stilling well. This was discovered when a craftsman removed the capacitance probe for inspection. He noticed that air was blowing out from around the threads as he unscrewed the probe. How could a vented tank build up pressure? Without vent holes, the rising level in the tank caused the stilling well vapors to be compressed. That's why at lower levels it was fairly accurate, but at higher levels it would lag behind.
THE CURE: Vent holes were drilled at the top of the stilling well to allow for proper level equalization.

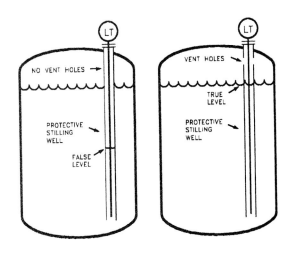

CASE #3

THE SYMPTOMS: A large process control valve (globe style) was suspected of being plugged. Since there was no flow meter on the line, I troubleshot it by using my hand to feel the line. The difference in the temperature upstream (hot) and downstream (cool) indicated that it was not passing the hot liquid. A x-ray of the valve in both the open and closed positions was requested. I was not given a chance to see the x-ray, but I was told that it was not plugged and looked normal. The line was taken out of service so that a water hose could be hooked to it for testing. An odd thing occurred. With the valve on the ground and stroked fully open, water could be made to flow in one direction through it, but not in the other. The valve was sent to the instrument shop for inspection and repair. However, it was found to be clean and had no indication of a problem. The shop foreman sent it back along with a scolding. He didn't want me to waste his time working on valves that didn't need repairing. After more testing and troubleshooting, the valve was still found to pass water in one direction but not in the other. Warning or not, I had to send it back to the shop. This time (What a relief!) they found the problem.

THE PROBLEM: The valve stem and plug were two pieces. The stem rested in the socket hole of the plug and was welded together like one piece. The welded bead had been turned down on a lathe. However, they had cut off virtually all of the metal from the weld. In time, the paper-thin weld broke. This allowed the plug to slide up and down on the valve stem when it was in the open position. That was why flow could be directed under the plug and it would pass the water. The plug would slide upward on the stem. And when the flow was directed over the plug, it would slide downward and block the flow somewhat like a check valve.

* By the way, a closer examination of the x-rays would have shown that the stem could move up and down within the socket, but the plug did not move.

THE CURE: A new one-piece valve trim was installed.

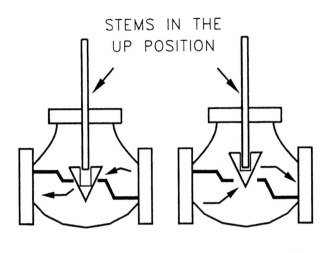

STEMS IN THE
UP POSITION

FLOW OVER
CHECKS FLOW

FLOW UNDER
ALLOWS FLOW

CASE #4

THE SYMPTOMS: Operations was suspicious that a pH transmitter was out of calibration and I was given the task. The range of the transmitter was 2 to 12 pH. I used buffer solutions of 4, 7 & 10 as the input standard and hooked a DMM (digital multi-meter) in series with the 4-20 ma output signal. At all three points the pH transmitter was barely off and only needed a minor zero adjustment. I thought to myself, "Another job well done."

THE PROBLEM: However, I needed to double-check my results, so I radioed into the control room. With the transmitter at mid point (the probe was in a buffer solution of 7) and the DMM reading 12 ma (50%), the DCS was indicating 40% of the range. For some reason there was a big difference in the readings. A check of the DCS point configuration confirmed that it had the correct range and signal conditioning. I went to the tech room and tried checking the ma signal (by placing my DMM in series) at the input terminals of the DCS. It was reading 10.4 ma which is 40%. Somehow the 12 ma signal sent by the pH transmitter was not getting to the DCS. I decided to use the "divide and conquer" method by checking the signal at the field junction box. I performed a current check there with my DMM and it too indicated a signal of 10.4 ma. Then I noticed something. The color of the signal wires was different at the junction box than it was leaving the pH meter. Was I checking the wrong wires? More investigation was needed. I removed the flex conduit on the pH meter trying to discover where the wires changed color. Rusty water poured out of the flex. With the conduit out of the way, I could see the problem. The wires had been butt-spliced. The dirty water, which had been trapped and lying in the conduit, had acted as a partial short circuit across the output signal wires. The water had been passing about 1.6 ma.

THE CURE: I ran new wires from the junction box to the pH meter and made sure that all of the flex was sealed tight to keep out the rain.

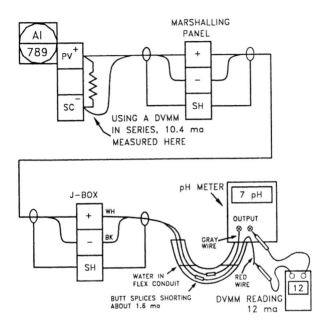

CASE #5

THE SYMPTOMS: A flow transmitter (4-20ma signal) was not reading correctly. Operations became suspicious when they noticed that the flow trend never got above 83%. When trended, the PV (process variable) had the typical ragged wandering signal, but flattened out smoothly at 83%. Something was effecting the signal.

THE PROBLEM: The flow transmitter was removed and calibrated in the shop. It all checked out perfectly. The primary flow device was a 90-degree weir, which used a bubbler dip tube to determine the flow rate. It too checked out perfectly. A loop simulator was used to test the signal into the control room. The two-wire loop was powered from the DCS. The signal would not go above about 83%. What was wrong? Again the calibration of the transmitter was checked. This time it was tested in the field. This time, it would not go above 83%. Originally this loop was a local pneumatic controller with a chart recorder. In other words a FRC. When it was upgraded and changed to be a DCS loop, they still wanted to use the recorder portion of the FRC. The problem was that the input resistance of the FRC was 500 ohms. By the time you add the 250 ohms for the DCS input plus some for the wire and you get a total loop resistance of about 780 ohms. There was too much loop resistance for a power supply of 24 VDC.

THE CURE: Since the DCS power supply of 24 VDC could not be changed and operations wanted to keep their circular chart, a local power supply of 32 VDC was installed. The loop was changed from DCS powered to be field-powered.

The 32 VDC would allow about 1000 Ω of loop resistance.

TOO MUCH LOOP RESISTANCE

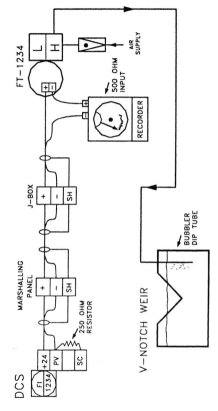

CASE #6

THE SYMPTOMS: A threaded ½ inch pressure-reducing regulator was checked on the bench and set at 55 PSIG. When installed in the process, it would not regulate at 55 PSIG or at any pressure for that matter. Apparently there wasn't any flow through it.

THE PROBLEM: The regulator was taken back to the bench, hooked as before and tested again. It worked perfectly on the bench. The process lines were checked for blockage by blowing pressure out of both the inlet and outlet lines. When reinstalled in the process, the regulator still didn't work. It was removed and examined closely. About ¾ of an inch down in the regulator's inlet, a small pressure port was noticed on the side. The process piping threads were extra long and tapered. When the regulator was screwed onto the process piping, the threads covered the small side port. However, when the tubing connectors

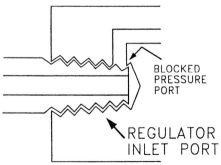

BLOCKED
PRESSURE
PORT

REGULATOR
INLET PORT

in the shop were screwed into the regulator for calibration and testing, they were of normal length and did not cover the small side port.

THE CURE: The process piping threads were cut off slightly and the regulator installed.

CASE #7

THE SYMPTOMS: A static head level transmitter had a history of being inaccurate periodically. No one could seem to find the problem. The level calculations were all checked. No matter how often it was checked, nothing was ever found to be wrong.

THE PROBLEM: Once when the taps were inspected, it was noted that the lower tap was connected at a right angle to the discharge pipe on the bottom of the vessel. It was a small diameter pipe. Could it be that the velocity of the process through the pipe was effecting the level measurement? As it turned out, the high velocity did tend to pull a vacuum on that line. Only during periods of low velocity did the level measure correctly.

THE CURE: The lower tap was changed to an alternate location. New calculations were made. The tap change was proven to be the correct solution because the periodic inaccuracies did not reoccur.

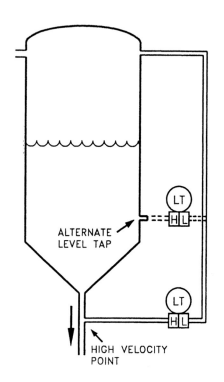

ALTERNATE
LEVEL TAP

LT

H L

LT

H L

HIGH VELOCITY
POINT

CASE #8
THE SYMPTOMS: One of the thermocouples (type K) in
the heater system was reading about 150 degrees instead
of the expected 500 degrees.

THE PROBLEM: Other temperature sensors nearby were reading around 500 degrees. So why was only one of the sensors reading differently? The screw-on lid of the thermocouple head was removed so that the sensor could be checked. A temperature calibrator was connected to the thermocouple and it indicated a reading of about 500 degrees. When a simulated signal was sent into the control room (DCS), it too indicated about 500 degrees. The thermocouple wires were reconnected and the thermocouple head lid was screwed back on. When the DCS was checked, it still indicated about 150 degrees. Again the thermocouple wires were checked. However, this time it was noticed that about three inches of the insulation on the type K thermocouple wire had been stripped off. When the thermocouple wires were stuffed into the head in order to close it, both wires (positive and negative) touched the metal screw-on lid. Thermocouple

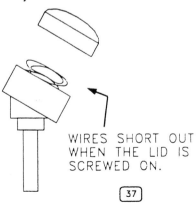

WIRES SHORT OUT
WHEN THE LID IS
SCREWED ON.

wires always read the last spot shorted out. In this case it read 150 degrees because that was the ambient temperature at the thermocouple head.

THE CURE: The excess bare wires were cut off. And to make sure that it didn't accidentally short out again, a rubber gasket was cut that snuggly fit into the screw-on lid. So when the lid was screwed on, the thermocouple wires would touch rubber instead of metal.

CASE #9

THE SYMPTOMS: The catalyst flow meters to train #1 and #2 had been fine for several years. Metering pumps were used to control the flow. But now, both of the flow readings would drift or indicate flow changes.

THE PROBLEM: During the troubleshooting effort, it was observed that disconnecting and reconnecting one of the flow transmitters (train #1) caused drastic changes to the reading on the other flow meter (train #2). The instrument craftsperson suspected a grounding problem. A common mode voltage check was made (see drawing below for details) and a grounding problem was revealed.

THE CURE: A loop isolator was installed in the field on both field transmitters. It's a good idea to customarily put a loop isolator on all self-powered transmitters that go to the DCS. You never know when changes in the grounding systems may create an excessive common mode voltage. The typical loop powered (2-wire) instruments do not need loop isolators.

COMMON MODE VOLTAGE CHECK

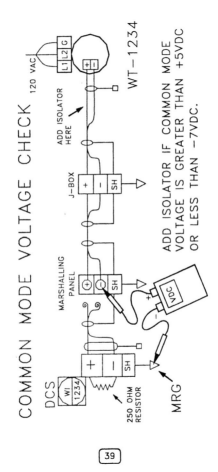

ADD ISOLATOR IF COMMON MODE VOLTAGE IS GREATER THAN +5VDC OR LESS THAN -7VDC.

TESTING FOR EXCESSIVE COMMON MODE VOLTAGES:

1. Disconnect the two instrument wires (both + & -) at the DCS terminal or Marshalling Panel. Have the DCS and the field transmitter powered up.
2. Use a voltmeter to measure the DC voltage in the technical room. Measure from the MRG (Master Reference Ground) at the DCS and the negative wire coming in from the field transmitter.
3. If the DC voltage is greater than +5VDC or less than −7VDC, you have an excessive common mode voltage and therefore a loop isolator is required.

CASE #10

THE SYMPTOMS: A control valve failed to work.

PROBLEM: The control valve actuator and its valve positioner was found to be loaded with water and therefore failed to work. I drained water out of fittings and let the air supply purge to atmosphere in order to get as much of the water out as possible. After it was dried out, it functioned perfectly. Just to be sure that it was repaired, I checked on it later in the day and found it to be squirting water out of the valve positioner once again. How could so much water be getting into the air supply? I traced the air line back to the instrument header and found that it was connected to a tap on the bottom of the pipe header instead of on top. This allowed the normal condensate that builds up in air supply lines to automatically drain down into the control valve.

CURE: I found an alternate location on the instrument air supply header that was on top of the header. This prevented the water from gravitating into the supply and down into the control valve. The problem did not reoccur.

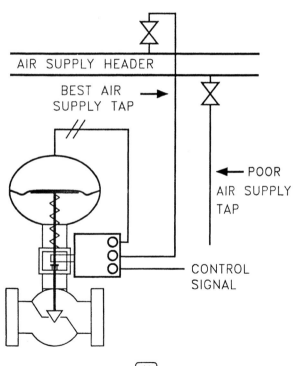

AIR SUPPLY HEADER

BEST AIR
SUPPLY TAP →

← POOR
AIR SUPPLY
TAP

CONTROL
SIGNAL

CASE #11

THE SYMPTOMS: The high level alarm on a compressor knock out pot was in question. The operators were worried that the instrument was not working. My job was to verify that the switch was working. I had to do all my testing while the knock out pot was in service.

THE PROBLEM: The level switch was the float style and it was installed in a chamber and bridle on the side of the knock out pot. See the sketch for details. The drain valve on the bottom of the chamber was a globe type and thus prevented me from using a rod to lift and test the float. It should have been designed with a ball or gate valve. By the way, even if I could have used a welding rod to lift the float through the drain valve, that would not have been an absolute test. If the float had a hole or crack, that would still keep it from working. I chose to test the float by adding water to the chamber and letting it actuate the switch. It would cause a little water to spill into the knock out pot, but that should not be a problem since the vessel was fairly large. With both process valves blocked, I hooked my utility water hose to the drain valve. I opened the top process valve and then cracked open the water hose valve. I could hear the water flowing. After a few minutes, the level switch had not given an alarm. Surely by now there was enough water in the chamber to lift the float, I thought.

THE SAFETY LESSON: I had overlooked the fact that the knock out pot pressure was 100 PSIG and the utility water was 60 PSIG. The flow that I was hearing was actually combustible hydrocarbon gases flowing **backwards** into the utility water system. I immediately shut off the water hose. This taught me a safety lesson. I should have had a check valve on the water hose and I should have considered the difference in pressures.

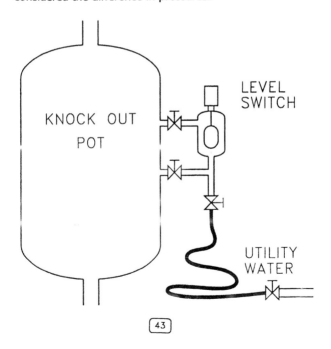

CASE #12

THE SYMPTOMS: I was asked to check out a differential flow meter and combination chart recorder. The steam flow meter was not reading correctly.

THE PROBLEM: The orifice was located up in the pipe rack. The impulse taps were run down to near ground level through a 3-valve instrument manifold. Below the differential instrument there were drain valves for both the high and low pressure connections. I blew down both the steam taps and they seemed to be clear. (Before blowing down, be sure that the manifold and valve seats can handle the hot steam temperatures.) However, steam can be very deceiving. I've found out from experience that you must blow down steam lines longer than you normally would. A 5-second blown is not long enough to verify that the taps are clear. By blowing down the taps for about 30 seconds, I saw that one of the taps was slowing way down. Finally it almost totally stopped. You wouldn't think that steam taps could get plugged, but they can. Scale had flaked off and fallen down into the 3-valve manifold. Once the tap was cleaned out, the instrument measured the correct flow.

THE CURE: The taps were revised so that scale could not fall directly into the 3-valve manifold or the instrument. A tubing tee was connected so that sediment would be trapped in a tubing tee.

CASE #13

THE SYMPTOMS: The indicator on a magnetic bridle level gauge was showing the wrong level on a propane vessel. The new magnetic bridle instrument had worked when it had first been installed, but now a few weeks later it indicated the level to be almost empty. It seemed to be unresponsive to level changes in the vessel.

THE PROBLEM: After quizzing the operators about operating conditions, it was determined that the operating pressure was now lower than normal. After checking the properties of propane, it was determined that there might be flashing in the bridle. When a liquid flashes and forms bubbles (boiling) the apparent specific gravity of that liquid is reduced. Because magnetic bridles are specific gravity dependent, the float will lose some of its buoyancy in the presence of flashing. The second problem was from steel filings suspended in the liquid propane. When the bridle was removed and examined, I found a fuzz ball around the magnet part of the float. Because magnetic bridles have close tolerances, the float had difficulty moving.

THE CURE: Three things were done to combat the flashing problem. First, the operating pressure was restored to the design value. Secondly, the magnetic bridle was fully insulated. And thirdly, a float designed for use in a lower specific gravity was employed. To help with the fuzz ball problem, a "magnetic trap" was installed in the inlet of the lower process connection. This kept the steel filings from reaching the float.

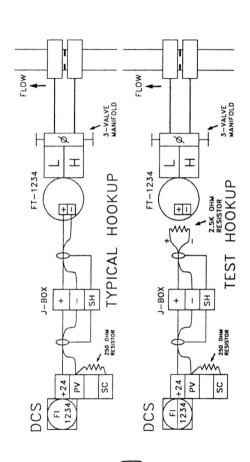

TYPICAL HOOKUP

TEST HOOKUP

☞ **SIMULATION EXAMPLE:** If an ordinary 4-20ma transmitter shows signal glitches or erratic behavior, the usual first thing to try is replacement of the transmitter. If that doesn't solve the problem, your next troubleshooting step could be to use a 2.5K ohm resistor to simulate a signal. See drawing on the previous page. A 2.5K ohm resistor and a 24 VDC power supply should give you a near mid-scale signal. If it doesn't, check your power supply voltage, the loop resistance or look for a shorted signal. The simulation resistor is used to replace the field transmitter. The resistor can be either connected inside the lid of the transmitter or in the field junction box. You can observe the 4-20ma signal by trending it with the DCS or by installing a recorder. If the simulation resistor gives you a very steady signal, the problem is probably in the process, taps, manifold or the transmitter. If you still have an erratic signal, the problem could be in the DCS input card or it could be induced electrical interference in the homerun cable or wiring. In one case it turned out to be caused by alkyls partially plugging the process taps.

COMMON INSTRUMENT CHECK OUT PROBLEMS

After a modification, repair or shutdown, there are usually a few problems with the instrumentation or controls. Listed below are some often-overlooked areas that may hinder your startup. It is always a good idea to make sure that construction crews and maintenance craftsmen are supplied with copies of drawings, specs, documents and vendor cut sheets to aid construction or maintenance craftsmen.

* PROCESS VERIFICATION: Whenever possible, verify the function of instrument loops and signals by using the actual process condition. Have operations apply temperature, flow, level, pressure, etc. to prove that it works. Do not totally rely on quick tests of the wiring or "shooting the loop".

* SHORT TEMPERATURE SENSOR: When a thermocouple or other sensor is too short and doesn't bottom out in the thermowell, there will be a big error in temperature measurement. All thermocouple, RTD and other thermal sensors should firmly touch the bottom of their thermowells.

* REGULATOR BACKWARDS: It is easy to install a regulator backwards. Not every manufacturer marks the inlet and outlet of their products clearly. Control valves can easily be installed backwards too.

* ENCLOSURE RATING: If a pressure switch has been replaced with a different brand or model, the new one may not meet the required area electrical classification.

* SEAL LEGS: Level transmitters may not have their seal legs fully filled thus causing improper indication. Steam siphons on pressure gauges or instruments will need to be filled prior to start up.

* CONTROLLER ACTION: When field controllers (usually pneumatic) are taken to the shop for alignment, often they are set for reverse action in order to check it. If their proper operating action is direct, and it doesn't get switched back, the controller will not work properly.

* SIGNAL WIRES CROSSED: One of the most common errors is to have signal wires crossed or wires on the wrong terminals. Check wires for continuity or have the wires phoned out to verify them.

* IMPULSE TUBING: The process tubing (impulse tubing) for an orifice flow meter could be crossed, therefore not showing a flow.

* DCS CONNECTIONS: Two-wire instrument loops are either field-powered or DCS powered (the most common). Make sure that you are connected to the proper DCS terminals. It is possible to have instruments hooked up so that both the DCS and the field device are supplying the power and it still works. However with two power supplies fighting each other, mystery signal anomalies can result.

* SIGNAL ISOLATORS: Field-powered DCS loops require loop signal isolators because of the common mode voltages. Without an isolator, experience has shown that signal drifting or spikes can result. Be careful when selecting "loop powered" isolators since some models can only handle input loop loads of no more than 275 Ω.

* LOOP RESISTANCE: If loop-powered transmitters (using 24 VDC) exceed a total loop resistance of 600 ohms, their output signals will top out before reaching 100%. This problem will not be detected if the transmitter is calibrated on the bench. Only a field calibration will reveal it. There are two ways to resolve it. You can reduce the loop resistance by eliminating devices within the loop or you can increase the loop voltage to possibly 32 VDC. Increasing the loop voltage will require changing the transmitter so that it is field-powered.

* MAGNETIC SWITCHES: Position switches that operate by magnetism, must have a target that can be attracted. Don't use SS or aluminum metals. Use a metal mounting bracket that doesn't interfere with the operation of the switch.

* WRONG RELAY: Since most relays look the same, it is easy to plug in the wrong type. A 24 VDC relay could be mistakenly used in place of a 120 VAC one or vice versa.

* WIRE LUGS: Make sure that the insulation has been properly stripped from the end of the wire. It is possible to tighten down the wire lug onto only the insulation and not have electrical continuity. Make sure that all of the terminals are tightened properly. If they are not snug, it is possible for large ambient temperature fluctuations to loosen lug terminals.

* FUSE VALUES: DCS systems need protection. Loops and controls that have fuses too big or too small can cause problems. Verify the correct fuse rating from the instrument loop drawings or from the assigned instrument engineer. Some digital output cards can handle a maximum current load of 2 amps. Typically the marshalling panel fuses will be the "fast blow" type so that they will blow before the equivalent rated DCS card fuses. Listed below are some typical DCS or ballpark fuse values for comparison.

VOLTAGE & TYPE:	FUSE SIZE:
24 VDC Analog Inputs	1/8 amp 250 Volt
24 VDC Analog Outputs	1/4 amp 250 Volt
120 VAC Digital Output	2.0 amp 250 Volt
24 VDC Digital Output	2.0 amp 250 Volt
120 VAC Digital Inputs	1/4 amp 250 Volt
24 VDC Digital Inputs	1/4 amp 250 Volt

Note: When sizing a power fuse that supplies several interposing relays, you must consider the amp load for each device. The total load for all the relays should exceed about 80% of the fuse rating.

* BLOWN FUSES: When fuses are installed before the wires have been checked and before the instruments have been installed and wired, they could easily blow from accidental touching of the wires.

* NO POWER/AIR SUPPLY: When loops don't respond during loop check out, make sure that they have their breakers and/or their air supplies turned on. Also check for missing fuses (or blown ones) and jumpers.

* LOGIC PREVAILS: If a field device does not function correctly with the DCS logic, resist the urge to change the logic. If the program logic has been thoroughly checked out, changes could introduce errors that could keep it from functioning properly. Instead, change the field device. The logic program should dictate the action and responses of all the field devices.

* CAN'T VIEW ALL OF THE LEVEL: Poor piping designs on level gauges could trap the level or interface level so that it will not follow along with the true internal level. For example, the upper process tap for an interface must be located in the upper liquid phase, instead of the vapor space. Otherwise it can't indicate the interface correctly. Below is an example of a trapped level gauge situation.

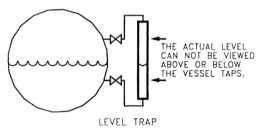

THE ACTUAL LEVEL
CAN NOT BE VIEWED
ABOVE OR BELOW
THE VESSEL TAPS.

LEVEL TRAP

* TOTALLY FAILSAFE: There are two failsafe scenarios for switches that are powered in the field. For example, take a temperature switch powered with 120 VAC that has a thermocouple input. The output of the switch is a form "C" set of dry contacts connected as a digital input to the DCS. Suppose you want the failsafe mode for the dry contacts to be open if the thermocouple opens. That's fine, but you also need to check what happens when the power is removed. With a loss of power, it could be that the dry contacts remain closed.

* MANIFOLD BACKWARDS: If three-valve instrument manifolds used in orifice flow applications are installed backwards, the D/P pressure transmitter can not be equalized and zeroed properly. Usually D/P transmitters can be calibrated in the field if they have the right manifold for it.

* MANIFOLD BY-PASSED: Three-valve manifolds should not be used in level service if a seal leg is required. Any accidental opening of the by-pass valve will dump part of the seal leg and cause a false level signal. Often 3-valve manifolds are used in flow orifice installations. In steam flow cases condensate forms a seal leg in the impulse taps and manifold. The condensate protects the manifold and the flow transmitter from the excessive heat of the steam. If the by-pass valve were opened along with the other two manifold valves, the seal leg of condensate would be quickly swept away. The directly applied hot steam could cause damage to soft manifold seats and/or the flow transmitter.

* POSITIONER BY-PASSED: When a valve positioner is used to reverse the action of a control valve or used as part of a split range control scheme, the positioner should not be by-passed. The loop can not function unless it is in the positioner mode.

* ORIFICE BACKWARDS: Orifice plates can easily be installed backwards and give low flow errors of varying magnitude. Standard square-edge orifices need the sharp edge and the printing on the handle facing UPSTREAM. (Some integral orifices are an exception to the rule.)

* NO FLOW: Check to see if the orifice plate was installed. Without an orifice the transmitter will not be able to measure any flow.

* CAPACITANCE SWITCH: When testing on/off capacitance switches on the bench, be sure that you use a valid method. Most capacitance switches can not be properly tested in a Styrofoam cup full of water. It may require that the container be conductive and grounded.

* MISS MATCH OF FIELD DEVICE: If the DCS configuration does not match the field device you will have a miss match. For example, if the limit switches on diverter valves are wired with one convention in mind (normally open) and the field device is wired under another convention (normally closed) they are miss matched.

* INPUT/OUTPUT PROBLEMS: When long runs (over 300 feet) are used on some DCS digital inputs, the induced voltage in the wire by other devices may cause it to mysteriously show to be closed. If shielded wire is used on the home run, the problem is resolved. Some DCS output signals have another problem. Sometimes when energized, they won't turn off. This problem has been traced to the RC network on the output card. Often RC networks are designed for use with loads greater than 100 milliamps. When light loads such as low power solenoids or relays are used, the leakage through the RC network is enough to keep the device energized. Two solutions have been used. Either use 24 VDC or install a larger interposing relay to actuate the light load.

* SHIPPING STOPS: All instruments need to be checked and/or calibrated prior to installation. Make sure that all of the shipping stops have been removed and that all of the dip switches or jumpers have been placed in their correct positions.

* REMOVE PROBE COVERS: Be sure that the vinyl end caps on pH sensors are removed before they are placed in service. Do not allow pH probes to dry out. On moisture probes, remember to remove the protective metal sleeve.

* BUTTERFLY VALVES: A "Fail Open" (F.O.) type butterfly valve will require an air supply to remove or to install it. Air must be applied to close the valve and give it a flat profile in order to slip between the flanges. Make sure that you have sufficient pressure to fully close butterfly valves, especially those with rubber liners/seats.

* WRONG SLOPE: When transmitters are mounted remotely, the proper slope is required on the connection (impulse) tubing. Dry legs may need to be able to drain condensate back into the process. Wet legs may need to be sloped in order to keep the seal fluid in place. Normally, the minimum tubing slope should be 1 inch per foot. Tubing that causes pocketing can affect the sensing accuracy of the instrument.

* WRONG REGULATOR PRESSURE: While some pneumatic instruments do operate on full line pressure, most require a specific supply pressure maintained by a pressure-reducing regulator. Insufficient supply pressure will keep it from working while excessive pressure may damage it. Check the instrument specifications to find out the requirements for pressure regulators, filters or lubricators.

TECH TIP: Don't fill a conduit over 40% full of wires.

* SKINNED WIRE: When too many wires are forced into a conduit, the insulation on the wires can be skinned and cause a wire fault. If a transmitter works on the bench but not when installed in the field, you may have a skinned wire going to ground. See **Wire Checks** below. For maximum conduit fill information consult the current NEC manual.

* **CHECK-OUT FUNDAMENTALS:** When brand new instrumentation is installed, several checks should be made. Among them are; wire checks, function checks, failsafe action verification and logic checks.

1) Wire Checks - The wiring should be checked for secure and correct connections, proper wire size, proper labeling, continuity, shorts to ground and excessive resistance. (Typically, short wire runs will have less than 4Ω, long runs will have less than 12Ω.) These checks should be done without the loop fuses installed.

High Voltage cables should be tested for faults.

2) Function Checks - (For Analog Inputs) After the transmitters have been calibrated and installed in the field, have them tagged. Install the loop fuses. Use actual process changes if possible or loop simulation devices to determine if the DCS receives the signal at 0%, 50% and 100% (check in both directions, increasing and decreasing). For analog outputs going to control valves, stroke the control valve from the DCS and see if it responds at 0%, 50% and 100% (in both directions, increasing and decreasing).

3) Failsafe Checks - In addition to the function test that manipulates (or simulates) the process to the alarm point, notice what happens when a wire is lifted.

* (For control valves) In turn, remove the 4-20 ma signal to the transducer, the air supply and the 3-15 psig air signal to the control valve or positioner to check for proper valve failure.

* Lift a one of the contact wires on simple process-powered (pressure, level, temperature, flow, etc.) switches to test how they will fail.

* Speed and temperature switches with electronics have two failsafe action possibilities, the failure upon power loss and failure because of an open contact or broken wire. First test for power failsafe action, by removing the power and seeing the results. Restore the power. Next, test the contacts by lifting a wire of the contacts and see the results.

4) Logic Checks - For Large interlock systems, test and simulate the logic and DCS display schematics beforehand by using another DCS system. For smaller or less complicated interlocks, force out points or use jumpers to verify their proper action.

* PRESSURE TESTING INSTRUMENT IMPULSE TUBING: The instrument tubing used to connect the process and the field instrument must be tested to see if the connections will hold at a specific pressure. It is dangerous to place it in service without testing the tubing.

* CONTROL VALVE ACTUATOR TOO TALL: Sometimes the piping for the valve drop is designed without checking the vertical dimensions of the control valve. Rework the pipe rather than let them install the valve at an angle.

* WRONG ACTUATOR PRESSURE: Many on/off actuators require at least 70 psig (or there about) and you may only have 60 psig of air header pressure available. Some actuators have a maximum pressure that should not be exceeded. Check to see the requirements. It is best to test valve actions on the bench before they are placed in the line.

* CORRECT SOFTWARE VERSION: If the wrong software version or revision is loaded, it may not work with newer files, printers or related programs.

* OUTPUT CARD JUMPERS: Certain DCS analog output cards, come with an output jumper. The DCS sends a tracer signal to check for output continuity. If it finds no continuity, a system alarm is generated. To prevent nuisance system failure alarms, jumpers are used across the outputs. Normally these jumpers are removed when the output point is configured and commissioned. Don't forget to check the jumpers if you have output problems during startup.

* MIXED PIPE SCHEDULE: If accuracy is important in an orifice flowmeter, don't assume that the meter run has been fabricated correctly. Check for different pipe thickness schedules up and downstream in the same meter run.

* PROCESS BLOCKED: Make sure that instruments are not blocked and sensors are open to the process. Also check to see that vent plugs are closed as well.

* ZERO METERS: Field transmitters should either be zeroed to the DCS system (my personal choice) or to an accurate DMM. Follow your own plant's policy about this. Some instruments need to be in their process environment in order to be zeroed properly. Magnetic flowmeters (magmeters) must be liquid full when they are zeroed. Some others that need the actual process for zeroing are some vortex meters, coriolis meters, ultrasonic meters and certain remote seal installations.

* TRANSMITTER STATIC PRESSURE SHIFT: It's rare these days with modern technology, but older differential pressure transmitters used on orifices and venturi flow meters can experience static pressure shift. Often they calibrate normally at atmospheric pressure, but when the full process pressure is applied, the output shows more flow (or less flow) than it should.

INTERMITTENT PROBLEMS

One of the most frustrating instrument or electrical problems to solve is the intermittent type. When a problem only shows up sporadically, that makes it very difficult to diagnose and repair. I have found three troubleshooting techniques that help with intermittent problems.

MAKE IT REOCCUR: When the problem occurs under a specific set of conditions, simulate those conditions to make the problem show itself. For example, if the problem is heat related, use a hair dryer to warm the circuit board so that the problem will occur. If the pesky problem is cold temperature related, use instant contact cleaner to simulate a cold condition. Often intermittent mechanical problems will show up when you introduce a small amount of vibrations or by tapping on it.

TRAPS AND INDICATORS: The more that you can sense about the problem, the better your chances are that you will solve it. Some problems can be trapped. You can use power line recorders (spike analyzers), oscilloscopes, or other devices to catch the problem in the act. Or you may need tattletale devices. Decide if you need to install temporary indicators, lights, buzzers, alarms, pressure gauges, voltage meters or signal meters on the equipment to aid you in your troubleshooting.

Some operators are reluctant to admit that they have abused plant equipment. By installing traps, you may find out that an instrument keeps failing because it gets steamed out every shift. One way to tell if a piece of equipment has been operated beyond its recommended temperature is to use monitoring labels that change color. These stick-on temperature labels make good tattletale devices especially if the operator is unaware of them.

REPLACE AND OBSERVE: A slower method is to replace suspected parts or circuit boards one at a time then observe the results. This methodical elimination process may take awhile. If you don't have the time, try replacing everything that you can at once. Mass replacements are not very cost effective however, but often they will result in a quick resolution to an intermittent problem.

INSTRUMENT FAILSAFE CONSIDERATIONS

DEFINITIONS:

Failsafe - The configuration or design of an instrument or device so that it fails to a safe condition or mode.

Shelf Condition - The state of the alarm contacts when the instrument is pulled off of the shelf. In other words either open (no continuity) or closed (continuity), with 14.7 psia and 60 degrees F applied.

Normally Closed (N.C.) - When referring to the alarm or annunciator, it means that there is electrical continuity from the field device or contacts and therefore it is not in the alarm state. When referring to the field contacts, it refers to their shelf condition.

Normally Open (N.O.) - When referring to the alarm or annunciator, it means that there is no electrical continuity from the field contacts and therefore it is not in the alarm state. When referring to the field contacts, it means their shelf condition.

Firesafe Valve - A control valve that will still maintain its ability to prevent seat leakage and function properly even with some exposure to a fire.

DEADMAN SWITCH: A simple failsafe device. Some lawnmowers can only be operated if a person manually grips the handlebar switch. If the operator suddenly dies, they will release their grip and the lawnmower will shut off.

FIRESAFE CONTROL VALVE: Control valves that are not firesafe may not be able to fail in the proper position due to mechanical fatigue or deterioration. A non-firesafe valve with soft seats will not be able to seat off properly.

FIRE SHUTDOWN DEVICE: A simple fire shutdown device can be made from 1/4" plastic tubing. The kind with ultraviolet sun light protection. If a fire breaks out, the tubing will melt. That will cause the air to leak off of the diaphragm of the shutdown valve, thus shutting down the flammable feedstock.

PICKING THE FAILSAFE ACTION: There are three failure positions that a control valve can have for spring/diaphragm control valves; closed, open or in place. It will require ancillary equipment to make a control valve fail in place. You must consider the ramifications of not only a local failure (just that valve) but a plant wide failure as well. If the action of a control valve is reversed, either via relays or a valve positioner, it will still fail according to the spring in the actuator.

FAILSAFE ACTION OF A SPRINGLESS CONTROL VALVE: Springless actuators can fail one of four different ways: open, closed, in place and undetermined. Piston actuators that use air pressure on both sides may not fail in a predictable position. A local air supply failure could cause it to fail one way and an air signal failure could make it fail another. Often a bias actuator spring is used to guarantee that the control valve will fail to the desired position.

CHOOSE DIRECT OVER INFERRED SHUTDOWN DEVICES: Interlocks and shutdown systems should use direct switches or contacts rather than inferred ones. Take the example of air conditioning flow needed in a small control room. The flow of fresh air is required to keep nitrogen leaks from accumulating. Pick a flow switch in the duct rather than the blower motor contacts. It is possible to still have the blower contactor pulled in and yet not have airflow into the room. The fan belt could come off or the duct be diverted or blocked in. Since the flow switch directly measures the airflow, it is the best choice.

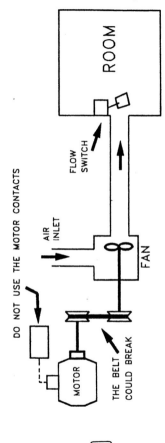

DO NOT USE THE MOTOR CONTACTS

AIR INLET

FLOW SWITCH

ROOM

FAN

THE BELT COULD BREAK

MOTOR

TECH TIP: When a new control loop works on manual but not in automatic, look at the controller's output signal. If it tends to drive the control valve either all the way open or all the way closed, it might have positive feedback. Change the action of your controller and see if that will solve the problem.

FAILSAFE HIGH AND LOW ALARMS: A failsafe alarm is one that has electrical continuity through the field contacts when the process condition is within typical operating parameters. The annunciator is satisfied and therefore does not sound the alarm. See the level switch example below. Note that the annunciator is connected to common and normally closed (N.C.) and the field device is also connected to the common and normally closed (N.C.) contacts. If either the wire gets pulled off or the field contacts open, the alarm will sound. All high state switches that are increasing (LSH, PSH, ASH, TSH or FSH) will be wired the same way. On the other hand, low state field switches that are decreasing (LSL, PSL, ASL, TSL or FSL) will be connected to common and normally open (N.O.).

FAILSAFE LEVEL SWITCH HIGH

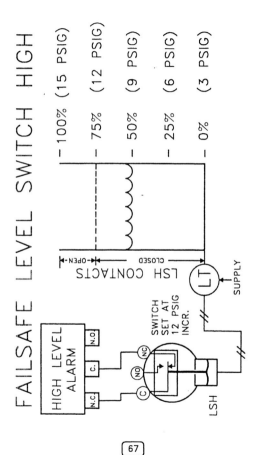

TEMPERATURE SWITCH FAILSAFE: Thermocouples can burn open. Some temperature transmitters or switches allow you to select upscale or downscale burn out. An upscale burnout would drive the output of a temperature transmitter to the maximum value. A temperature switch with upscale burn out will respond as if it has a high temperature. Usually temperature transmitters on heaters need upscale burnout so that the temperature controller will cut back on the heat, thus preventing the heater from burning up.

Besides the burn out issue, there are two other failsafe considerations for certain types of temperature switches; the shelf condition of the contacts and contact changes due to loss of power. See the example below. If the contacts are held in a certain position when the instrument is powered up, loss of power will make them change state. A true failsafe mode for a temperature switch is a loss of continuity for both a power loss and an open contact.

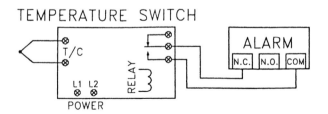

TEMPERATURE SWITCH

MUD DAUBERS AND SPIDERS: Control valves and solenoids may not fail properly if mud daubers or spiders block exhaust ports. Use bug screens to prevent bugs from making their homes in open ports.

TESTING FOR FAILSAFE ACTION: The most common way to test a failsafe device is to actually remove the signal, remove the power or to lift a wire.

POWER RESTORED: Not only must loss of a signal or power be considered, but also what happens when the power is restored. Flame outs on furnaces or boilers are dangerous. If the signal to a gas burner is lost, it should be designed so that it must be manually reset before it can be used again.

FAILSAFE ACTION WITH A PLC: Since PLC's have so many program choices, you must be careful when selecting the proper action. Usually you can configure it so that in the event of a processor card failure, the outputs de-energize. To avoid any confusion, take extra time to verify that your program is truly failsafe. Use plenty of remarks within your program to clarify it.

TRICKS OF THE TRADE
INTERMITTENT ELECTRICAL PROBLEMS

One way to find the problem when you have intermittent compressor shutdowns is to temporarily use clip-on fuses across the contacts in the shutdown circuit. The value of the fuses needs to be rated for about 10% of the current draw of the contactor or relay used in the circuit. Don't use a fuse rated over 70% of the current. Once the equipment is powered up and running, add the fuses across each switch in the circuit. Be careful when doing this since you will be working with a live circuit. As soon as you experience a phantom shutdown, check each of the fuses. The blown fuse will indicate the switch which opened first thus causing the shutdown.

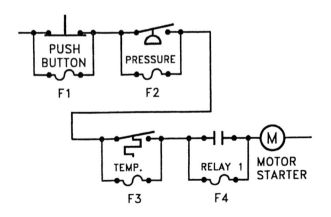

TESTING A 0-5 AMP CURRENT TRANSMITTER

Use three light bulbs (200 Watt each) in parallel as the load when testing a current transmitter with a range of 0-5 Amps. Use a Powerstat (auto-transformer) to vary the input amperage. Some of the older model transmitters may need a 250 Ω load on the output signal, but most of the modern ones will not.

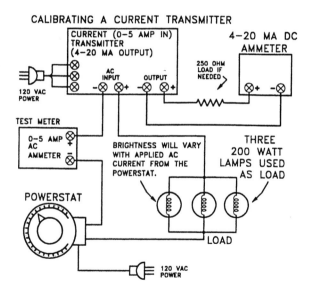

CALIBRATING A CURRENT TRANSMITTER

TECH TIP: If you need to choose an electrical enclosure for outside use in a Class 1, Division 2 area, you may need an enclosure that has a NEMA 4 (to keep out moisture) and a NEMA 7 (to contain internal explosions) rating. Some brands may offer one or the other rating, but not both. If that's the case, always choose NEMA 7 over the NEMA 4 and hope that moisture isn't a problem.

ELECTRICAL DEAD SHORT TEST

When many wires and devices are involved in dead short, you may be able to use a timer and a bulb to quickly find it. Hook up a repeat cycle timer (Eagle #CG102A6) set for a 2 second cycle and a 60-watt bulb as shown across the blown fuse. The bulb will reduce the short circuit current to about .5 Amps. Use a clamp-on amp meter to locate the pulsating current. You can clamp around bundles of wires and test them. Once you have found the 2-second pulsation in a bundle, divide it in half and check again. By the process of elimination, you can narrow down the problem until you have found the shorted wire or device.

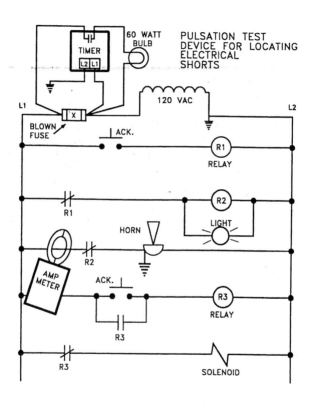

PULSATION TEST DEVICE FOR LOCATING ELECTRICAL SHORTS

60 WATT BULB

TIMER

L2 L1

120 VAC

L1

L2

BLOWN FUSE

X

ACK.

R1
RELAY

R1

R2
RELAY

LIGHT

HORN

R2

AMP METER

ACK.

R3
RELAY

R3

R3

SOLENOID

73

MEASURING A FUSE WHILE IN SERVICE

When working with live circuits, always do so safely. If you need to check a fuse in a circuit, use a voltmeter that can safely measure the expected voltage. One good idea is to begin by using the highest range. Another good idea is to use the "one hand" method so a not to provide a current path through the heart area. When you measure across the fuse (points A & B in the drawing), if you get the expected line voltage, the fuse is open. However, if you get no voltage, you will need to check further since there may not be any voltage at the fuse. Next measure from one side of the fuse (point A) to a known neutral (point C) and from the other side of the fuse (point B) to a known neutral (point C). If you find line voltage on both sides of the fuse to the neutral, the fuse is good. If the measurements are intermittent or in any way odd, turn off the power and investigate further.

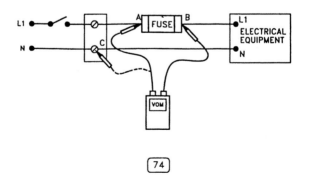

SOLENOID LEAKAGE CHECK

Notice the 4-way solenoid and piston actuator shown in the drawing below. The solenoid exchanges the air supply pressure with the exhaust in order to open and close the on/off actuator. When air is found leaking from one of the exhaust ports, usually the solenoid gets the blame. But about a third of the time, that's not the problem. A faulty o-ring in the piston actuator could be the problem. If the solenoid is in the position shown in the drawing, the supply air flows through the solenoid, then out line #1, then into the actuator. The other side of the piston vents off its pressure through line #2 and out the exhaust port of the solenoid. If you remove line #2 (be careful with air pressure) and air is blowing from the actuator, the problem is in the actuator. If air is blowing from the solenoid, then the problem is in the solenoid.

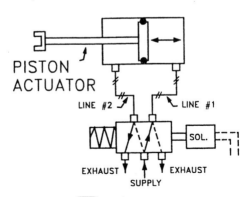

PISTON
ACTUATOR

LINE #2   LINE #1

SOL.

EXHAUST   EXHAUST
SUPPLY

POSSIBLE FIXES FOR REOCCURRING PROBLEMS:

- Pneumatic signals and (some process pressures) with noise or wild oscillations can be dampened several ways. You can try porous metal snubbers, volume pots, precision needle valves or an inverse derivative relay (Made by Moore Products).

- Taps that keep plugging may need, purge relays installed, a back flush system operated by a timer, a "rod out" device installed, the transmitter to be replaced with a diaphragm seal type or dual taps with a selector valve.

- You may need to guard against excessive pressure surges on delicate pressure instruments or analyzer sample systems. You can try a "lute" (dip leg), a relief valve, a rupture disk, a back-pressure regulator or an "over-pressure" lockup device.

- Spiders, bugs and mud daubers like to build their nests in vent holes, exhaust ports and solenoid openings. Install "bug screen" devices to prevent bugs from negating safety devices.

- To guard against excessive flow rates you can try using a restriction orifice, a flow regulator or a flow fuse.

Rules of Thumb:

➢ How hot will the instrument impulse lines (taps) get? You lose about 100 degrees of heat for every foot of non-flowing impulse line.

➢ Choose linear valve trim if the differential across the valve stays virtually constant. Choose an equal percentage trim if the differential pressure across the valve decreases as the valve opens. When in doubt, select an equal percentage trim.

➢ Depending upon the type of valve and its rangeability (turndown), most control valves become unstable and won't control very well below 10% of their flow capacity.

➢ Try to select a control valve trim so that the normal flow falls within 25% and 75%. The ideal normal flow would be at 60% of the trim's capacity.

➢ Most orifice flowmeters have a rangeability (turndown) of about 4:1. Don't expect accurate flow indications below 25% of the scale.

➢ When installing orifice flow meters you should have at least **20** pipe diameters of straight pipe upstream and at least **5** pipe diameters of straight pipe downstream. To be safe, you should check with the manufacturer since other factors such as the beta ratio could mean that even more straight pipe diameters are necessary. (The term "straight pipe" means no elbows, block valves, thermowells, pipe restrictions, sample taps, pressure gauges or other intrusions that could cause turbulence.)

- ➤ The best arrangement for an orifice flow loop is to place the orifice upstream and the control valve downstream. There are two main reasons for this arrangement. First, the orifice pressure will tend to be steadier since it rides on the supply or header pressure. This will help keep it at the design pressure for better measuring accuracy. Secondly, since modulating valves cause turbulence, it is better to have them downstream of the orifice. This will also help the measuring accuracy.

- ➤ Self-operated regulators have much quicker response over a typical pressure control loop containing a pressure transmitter, a transducer, a control valve and a controller.

- ➤ Some rotameter floats are difficult to read because they are odd shaped. Where do you read? Is it the top ring, the sharp edge, the beveled ring or what? One quick way to find out the reading point is to momentarily shut off the flow. With no flow you can readily see what lines up with the zero on the scale.

- ➤ When two feedstock fluids or utilities come together, you must guard against cross contamination. For example take a gas treater which normally uses a hydrocarbon gas. During the regeneration process, it must be switched over to use nitrogen gas. Use either physical separation (the safest choice), a spectacle blind or "double block and bleed" to keep the higher pressured gas from flowing into the lower pressured gas.

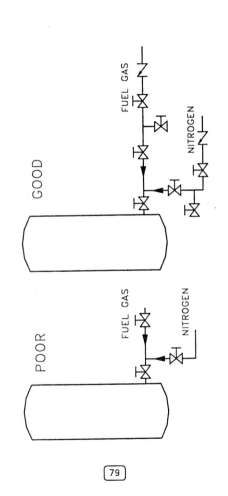

POOR

GOOD

FUEL GAS

NITROGEN

FUEL GAS

NITROGEN

- With all things being about equal, buy a packaged control system rather than piecing the parts together yourself. When complex control systems require troubleshooting, you will be on your own if you buy several pieces of vendor's equipment. Each vendor representative will want to work on their equipment, no the entire system.

- When designing "Speed Loops" for a slip-stream sample rotameter, use a 10:1 ratio. In other words, the sample rotameter flow would be 1/10 of the by-pass rotameter flow.

- Use a loop isolator when hooking field-powered transmitters to a DCS system. This will help you avoid signal drift due to ground loops.

- Leaks never get better. Temporarily they may appear to be improving, but in the long run they always get worse. Tag all leaks and fix them during shutdowns.

- Never reuse the electrical tape that was used to insulate the "T" leads of a motor connection. When disconnecting the "T" leads of a motor, you can cut it open by slicing down one side. This forms a hinge and allows it to be opened up like a clam for removal. When reconnecting the "T" leads, there is a temptation to save time by simply placing the "clam" over the connection and taping it closed. Don't take that short cut because the rubber tape and glass tape will be compromised. There is a real chance that the connections will short out eventually.

"T" LEADS
TAPED UP→

MOTOR

WRONG!

➢ When blending (converging) two gas streams with a 3-way valve, the Cv should be selected so that it makes up at least 80% of the system pressure drop. Anything less will cause it to act almost on/off instead of blending the two streams.

➢ No valve is "Bubble Tight". All valves can leak. Regardless of the type: block valves, solenoid valves, control valves, soft seated valves, check valves or what ever, do not assume that the valve will hold perfectly. Three-way and double-ported control valves are terrible about shutting off.

➢ About 75% of all "temporary" alterations or modifications somehow become permanent. You should design and install all temporary installations as thoroughly as if they were to be permanent.

➢ Run conduit in a manner to keep moisture from draining into an instrument. Make sure that drains are properly located. Do not run conduit next to steam lines.

➢ An air to open (ATO, or A/O) control valve will always fail closed and an air to close (ATC, or A/C) control valve will fail open unless there is a reverse acting positioner or some other control equipment that makes them do otherwise.

➢ The test equipment (your calibration standard) should be at least four times more accurate than the instrument under test.

FINDING CHLORINE LEAKS

Chlorinators and other devices that contain chlorine can develop leaks. One quick way to spot the source of the leak is with ammonia. Use a plastic bottle as shown below half filled with household liquid ammonia. Remove the plastic dip tube from inside the bottle. When the bottle is squeezed, only the ammonia vapors will be expelled. When the vapors encounter chlorine gas, a white cloud is momentarily formed to let you know that you've found the source of the leak. As with any potentially dangerous chemical, always handle it with safety in mind. Be sure to store the ammonia in a safe place. Check with your safety department to make sure that it is okay to use ammonia for detecting leaks in your plant or facility.

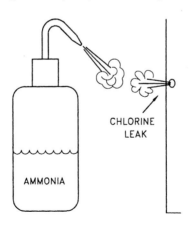

CHLORINE
LEAK

AMMONIA

UNDERSTANDING 4-20MA OUTPUTS

Most field transmitters are 2-wire loop-powered devices. The output from a controller uses 2 wires too. However, they are different. The controller uses a variable DC voltage to push its 4-20 MA signal. It could be just about any voltage up to 24VDC depending upon the output signal and the loop resistance. Notice the typical output shown below. The input resistance of the I/P transducer is 176Ω and the wires have 4Ω of resistance. Making a total of 180Ω for the loop. If the output from the controller or DCS is 8MA (25%), what voltage would you expect to measure across the transducer?

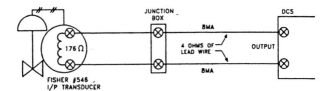

Using Ohm's Law, we see that it only uses 1.408 VDC. The controller generates only enough voltage to push the desired milliamp output. Below are some variations of Ohm's Law. Use the right hand oval for this calculation.

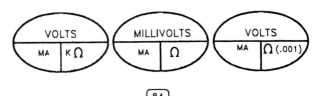

The formula becomes: $(MA)(\Omega)(.001) = $ Volts

(8MA) times (180Ω) times (.001) equals 1.44 volts

BUCKET METHOD FLOW TEST

One way to verify a flow is by using the "Bucket" method. I once used it on magnetic flow meters in a caustic soda slurry service. The normal calibration method was not possible because the electrodes developed a resistive coating from the caustic soda. Since the amount of coating kept building up, the only viable method for calibrating the magnetic flow meters was to use the "bucket" method. It required a platform scale, watch, a hose and a bucket. See the drawing below.

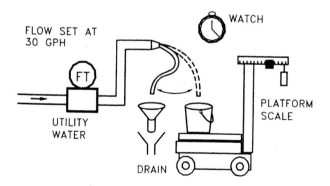

The flow test in my case was at a rate of 30 GPH. 30 GPH was the typical operating set point for these low range magnetic flow meters. For the test, I used utility water in place of the caustic soda. Utility water was selected because fluids measured by magnetic flow meters must have a sufficient amount of conductivity. Do not make the mistake of using distilled water because it lacks enough electrical conductivity. Once the water flow through the magnetic flow meter had been stabilized at 30 GPH, I diverted the water flow into the bucket for exactly 3 minutes. I weighed the water (remember that you must subtract the weight of the bucket) on the platform scale. I multiplied the water weight (in pounds) by a factor of 2.4 to get the gallons per hour (GPH). By the way, a factor of .04 would be used if you needed to convert it to gallons per minute (GPM). For example, if the weight of the water turned out to be 12 pounds, the calculated flow rate would be 28.8 GPH. When I found a difference between the flow shown on the controller and the "bucket" rate check, I adjusted the magnetic flow meter to match the "bucket" flow test results. I found that the "bucket" method typically gave me an accuracy of about +/- 2% to 3%.

POWER STRIPS

When you need to build a power strip, daisy chain it the smart way. See the example on the left below. If the power needs to be added or disconnected with the circuit live, it could be done without disrupting the power to the other points. For example, if point C needs to be disconnected, it can be done without points D or E losing power. If you had planned ahead for future expansion, you could safely add points F and G. Even if you also needed to add point H later, it would only affect G directly above it. Notice the other daisy chain method on the right below. It is set up poorly. If you try to disconnect point C, you will probably lose the power to D and E below it. Adding new points with the circuit live would be very dangerous. Besides the safety issue, it gets pretty crowded with three wires under each screw.

```
CORRECT          INCORRECT
DAISY CHAIN      DAISY CHAIN
METHOD           METHOD
```

HOMEMADE TEST EQUIPMENT

You can easily make test equipment yourself. A ballpark calibrator can be made from a regulator, a receiver gauge and a small block valve. It can be used to stroke pneumatic valves with 3-15PSIG actuators or to be an input signal for pressure transmitter or switch. By blocking in the valve you can use it to measure 3-15PSIG output signals.

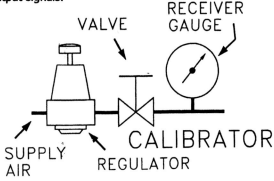

You can build a valve stroker from a few electrical parts. See the drawing below. It will allow you to send a 4-20MA current signal to the I/P transducer and thus actuate the control valve. Since it is not properly rated, you will not be able to use it in hazardous electrical areas.

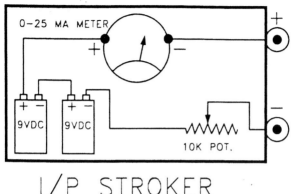

I/P STROKER

USING RELAYS TO TROUBLESHOOT
Relays can be used to help you troubleshoot circuits. If your relays don't have the neon light indication, change them out so they do. It only costs a little more to have that feature and it alone can save you a lot of troubleshooting time. Relays can help you to turn on or turn off outputs by merely pulling them out or plugging them into their sockets.

WEIRD FUSE CASE
I once traced the lack of power on a piece of electrical equipment to an apparent blown fuse. In the circuit it was tested open and when taken out for testing, it was found to have continuity. I found that an end cap was improperly soldered. It was an intermittent fuse.

TROUBLESHOOTING QUIZ

What could cause this thermocouple temperature transmitter to read lower than it should?

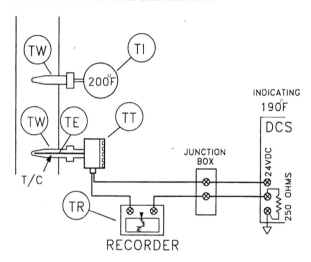

POSSIBLE CAUSES:

1. Loose wire or connection on the thermocouple (T/C) thus causing resistance.
2. The metal cover on the T/C head is shorting out the T/C wires.
3. Wrong type of T/C.

4. The thermowell is coated.
5. The T/C is not bottomed out in the thermowell.
6. The transmitter range and the DCS range are configured differently.
7. There is moisture in the junction box causing a partial short in the 4-20MA output signal wires.
8. Bad T/C sensor or temperature transmitter.
9. The T/C wires are connected backwards.
10. The temperature indicator is faulty.
11. There is too much loop resistance (exceeding 600 ohms).
12. There is a problem with the DCS. (very rare cases)

In the level case below, the level in the glass and the indication on the DCS are different. What could cause them to disagree?
1. One of the A or B valves are blocked.
2. The seal leg fluid is low or gone.
3. There is a transmitter zero shift and the calibration is off.
4. The range is configured wrong on the DCS.
5. Wrong initial calculation and calibration on LT-123.
6. Partial short of the wiring in the junction box.
7. The process specific gravity has changed.
8. The equalizing line on LT-123 is plugged.
9. The transmitter was zeroed according to the level glass instead of according to the transmitter taps.
10. The transmitter is malfunctioning.

11. One of the C or D valves on the level glass are blocked.
12. The level glass taps are plugged.
13. The process level has gone above the B valve and therefore displaced the .5 specific gravity seal leg with the heavier .65 process fluid.

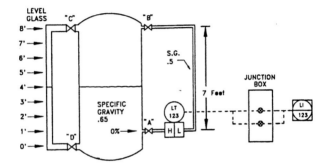

PRE-STARTUP CHECKLIST
LABELS/TAGS
- Hand switches (all types)
- Instrument Loop Tags
- Junction Boxes
- Warning Labels ("Danger High Voltage" tags)
- Internal terminal strips and wires
- Relays, solenoids, seal fluid ID

POWER
- Was the "Smoke" test successful
- Air supply, regulator, filter, lubricator turned on
- 120 VAC power turned on, fuses and jumpers installed
- Are the power disconnect switches lockable

VALVES/REGULATORS
- Oriented in the correct directions
- Regulators set to proper pressure
- Tubing properly connected to positioner, I/P, etc.
- All pipe flange bolts tight

MOUNTING/BOLTS
- All bolts tightened on explosion proof boxes
- Instruments securely mounted
- Instrument accessible, cover not blocked

WIRES/CONDUIT
- Conduit seals properly filled
- All conduit covers on elbows
- All grounding cables in place
- Wires and conduit properly hooked to instruments

MISCELLANEOUS
- All enclosures are suitable for the area classification

- Enough block valves for air supplies, instruments
- Thermocouples bottomed out
- Instrument manifolds (3 valve) installed properly
- Instrument seal legs filled
- Orifices not installed backwards
- Proper up and downstream pipe diameters for flowmeters
- PSV's not blocked in
- Have purges and rotameter flows been set
- Freeze precautions taken
- All shipping stops removed

TECH TIP: Before measuring an unknown AC voltage with your DMM (Digital Multi-Meter), verify that your test leads are in good condition. One way is to first measure for continuity with your DMM set on the ohms scale. With that successfully done, switch your DMM to the AC voltage scale and measure a known voltage, maybe a wall outlet. Now you can try measuring the unknown voltage. If your DMM doesn't show any voltage, test the leads again by measuring the known AC voltage again.

GENERAL INSTRUMENT SAFETY TIPS:

- Before opening a process transmitter, check the wind direction and identify your escape route. You may also need to locate a fire extinguisher and an eye wash station. If there is a danger of personnel getting sprayed or dripped on, rope off area. Always bleed or drain transmitters away from you.
- Never try to install packing in a control valve while it is under process pressure. If the packing gets pressured out, you could be injured or sprayed with process fluids.
- Be familiar with the plant fire and first aid phone numbers.
- Since the oxygen could get displaced, don't enter tanks purged with nitrogen or enclosures that could trap gases.
- Always use at least two methods to verify that you are working on the correct instrument. You could have an operator identify it, you could use a HART communicator, you could check the instrument loop metal tag or you could use the P&ID to help trace it out.

TECH TIP: For safety, when working around high voltages or in a Motor Control Center (MCC), never work alone.

- Always wear the appropriate personal safety equipment when working on hazardous processes, noisy equipment or near high voltages. You may need rubber gloves, goggles, hard hat, respirator, earplugs, steel-toed shoes, flame-retardant clothing or other safety items.
- Most controllers need to be placed in manual before you can work on them. Determine if interlocks or shutdowns could be tripped by you working on an instrument.
- Follow your company's Lock, Tag and Try safety procedure.

TECH TIP: Antifreeze (ethylene glycol) which is often used as a seal fluid has a specific gravity of 1.125 at full strength (100%) and 1.06 when dilute to 50% with water. 50% antifreeze will not freeze until it gets down to about −35°F. It's a good idea to use colored antifreeze in your instruments so that you will know if it is present or not. Be aware that antifreeze is poisonous when swallowed.

INSTRUMENT SAFETY TIPS IN THE FIELD

1. Always notify operations personnel before you touch a piece of equipment in the field. And NEVER initiate a process change (rate change, set-point change, valve closure/opening, equipment start/stop). LET THE OPERATOR DO THIS FOR YOU. If an accident or process problem occurs, you are covered. Otherwise, those problems will be YOUR FAULT.

2. When you are trying to determine if the instrument is under pressure, stand up wind. Rod it out or test it yourself. Do not assume that an instrument is depressurized, or even accept another person's word that it is depressurized.

3. Do not bypass a steam service orifice flowmeter. It will sweep out the protective cool condensate and cook the seats of the 3-valve manifold and the instrument.

4. Do not drain fluids above walkways without roping off the area.

5. Do not attempt to service a valve or other instrument while it is in service. This includes the replacement of packing to a valve in service and under pressure. This is a violation of the Lock, Tag & Try procedure.

6. If you try to hook a HART communicator onto certain brands of smart transmitters without putting the controller in manual first, its signal will be disrupted. Other smart transmitters will allow polling, but you must place them in manual before making any

changes. If you are only putting the controller in manual, then alarms, interlocks, other tags, or programs looking at that PV will respond to the changes in that PV. If you place both the controller and the PV in manual, then you should be pretty safe, but not 100%. If any other programs, tags, or interlocks are looking at the PVRAW parameter on the loop you are working on, then you have the same problem. There is no reason why anyone would use the PVRAW parameter, but that doesn't mean that somebody hasn't done it. The safest and only 100% sure-fire way to guarantee calibrating a loop won't effect anything else is to check with the DCS group.

7. As a rule of thumb never apply more than 100 PSIG on pristine plastic ¼" tubing.

8. Do not try to carry heavy instruments while climbing up a ladder cage. You need both hands for climbing. Use a tag line and/or a bucket for transporting items.

9. It's not a good idea to "dust yourself off" by using instrument air, utility air, or nitrogen. The level of pressure present in these sources can "inject" the dust, rust, chemicals, gas, or other foreign material into the body via epidermal permeation.

10. Verify twice before removing or lifting wires since it could be mislabeled. Remember that marshalling panel terminals can be confusing (parallax effect) or the numbers could have shifted up or down on the track. When tracing out wires, watch for changes in

color or size since that may indicate that you have gotten on the wrong wires accidentally.

11. Never break the circuit on current transformers (CT's) while the circuit being measured is still energized. This can cause the CT to overload.

12. Each DMM (Digital Multi-Meters) should be checked and sent out for calibration once a year to make sure that it is in good working order.

13. Be careful that you have the proper setting on your DMM before connecting to a circuit. If you want to measure a voltage, but you have it set for current, it will short out the circuit. If using an analog meter that is non-auto ranging, use the largest range first and then step down until you can measure the circuit near the middle of the scale.

14. Except for loops that you are checking out, do not silence alarms or reset equipment when an operator is not present.

15. Do not wear FRC (Fire Retardant Clothing) when you are working inside a DCS cabinet. Some FRC can build up static electricity charges and damage delicate IC (integrated circuits) chips.

16. Do not over tighten the packing on a leaking control valve stem. You could possibly bind it so that it can't stroke properly.

17. If a ball or plug cock valve is taken out of service while closed, the trim could have trapped process pressure inside. Be careful when it is opened.

1 to 5 VDC Simulator

Many transmitters have a 4-20ma input signal. But sometimes you will need to simulate a 1 to 5 VDC sensor signal to a transmitter or process indicator. For example take a Red Lion model PAXP0000 or similar instrument. Suppose it has an input range of 0-250 psig. 1 VDC represents 0 psig and 5VDC represents 250 psig. Of course you can use a pump-up pressure calibrator to apply pressure to the sensor for calibration of the instrument. After it has been calibrated, there may be times you need to test its action by applying a 1 to 5 VDC signal. Here is a simple way to simulate it. All you need is a 250Ω resistor and a 10KΩ potentiometer. Hook them up as shown in the drawing. If the Red Lion has been calibrated along with the pressure sensor previously, you can read the simulated input from the front panel of the Red Lion. Or another way is to use a DVM to measure across the 1 to 5 DC voltage applied at the input terminals (Volts Input and Common).

So by rotating the potentiometer knob left and right, you can precisely ramp the signal up and down to see where the alarm contacts open and close.

1 TO 5VDC SIMULATOR
FOR RED LION PAXP0000

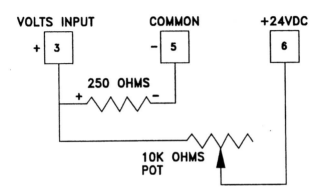

• •

WRONG FUSE PLACEMENT
The two electrical drawings below show two possible
locations for the loop power fuses. The first one is
the wrong location. Any short in the field would take
out the power panel and other loops, not just that one
loop. The second drawing has it in the proper
location. In case of a field electrical short circuit, only
that one loop would be taken out.

POOR FUSE PLACEMENT

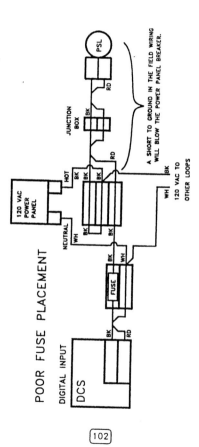

DIGITAL INPUT

DCS

FUSE

BK

RD

BK

WH

NEUTRAL

WH

BK BK

120 VAC
POWER
PANEL

HOT

BK

BK BK

WH

120 VAC TO
OTHER LOOPS

BK

RD

JUNCTION
BOX

BK

RD

PSL

A SHORT TO GROUND IN THE FIELD WIRING
WILL BLOW THE POWER PANEL BREAKER.

PROPER FUSE PLACEMENT

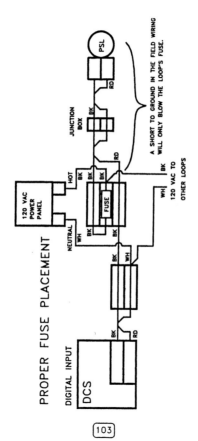

PSL

JUNCTION BOX

A SHORT TO GROUND IN THE FIELD WIRING WILL ONLY BLOW THE LOOP'S FUSE.

RD

BK

RD

HOT

BK
BK
BK

120 VAC POWER PANEL

FUSE

BK

120 VAC TO OTHER LOOPS

WH

NEUTRAL

WH

BK
BK

BK
WH

BK

RD

DIGITAL INPUT

DCS

TABLE OF CONTENTS

If you liked this book, you may also like to purchase
Purdy's Instrument Handbook, which was also written
by Ralph Dewey. It too is available from Glen Enterprises.